Plants of the Redwood Forest

Alice deMont

Copyright © 2012 by Alice deMont
Photographs and text by Alice deMont

All rights reserved. No part of this publication may be reproduced or transmitted in any form or by any means, electronic or mechanical, including photocopying, recording, or by any information storage and retrieval system, without prior permission in writing from the author.

All the effort has been made to verify the information in this book. However misidentification is still possible. For the confirmation concerning edibility and medicinal use of the plants consult the appropriate sources. The author does not accept any responsibility resulted from the use of the information presented in this book.

ISBN-13: 978-1480135093
ISBN-10: 1480135097

If you would like to order additional copies of this book, go to the following web-site: https://www.createspace.com/4031136

Table of Contents

Introduction 5
Cupressaceae 6
    Coast redwood (Sequoia sempervirens) 6
Oxalidaceae 7
    Redwood sorrel (Oxalis oregana) 7
Montiaceae 9
    Candy flower (Claytonia sibirica) 9
Asteraceae 10
    Coltsfoot (Petasites frigidus) 10
    White hawkweed (Hieracium albiflorum) 11
    Pathfinder (Adenocaulon bicolor) 12
    Nipplewort (Lapsana communis) 13
    Common madia (Madia elegans) 14
Brassicaceae 15
    Milkmaids (Cardamine californica) 15
    Hairy bittercress (Cardamine hirsuta) 16
Lamiaceae 17
    Betony (Stachys) 17
    Common self-heal (Prunella vulgaris) 18
    Lemon balm (Melissa officinalis) 19
    Field mint (Mentha arvensis) 20
Fabaceae 22
    Lesser hop trefoil (Trifolium dubium) 22
    Sweet pea (Lathyrus odoratus) 23
Rosaceae 24
    Thimbleberry (Rubus parviflorus) 24
    Himalayan blackberry (Rubus armeniacus) 26
    Salmonberry (Rubus spectabilis) 28
Berberidaceae 29
    Deerfoot (Achlys triphylla) 29
    Inside-out flower (Vancouveria hexandra) 30
    Oregon grape (Mahonia aquifolium) 31
Apiaceae 32
    Pacific sanicle (Sanicula crassicaulis) 32
    Water parsley (Oenanthe sarmentosa) 33
Grossulariaceae 34
    Currant (Ribes) 34
Ericaceae 35
    Huckleberry (Vaccinium) 35
Saxifragaceae 36
    Alumroot (Heuchera) 36
    Piggyback plant (Tolmiea menziesii) 37
Dryopteridaceae 38

  Lady fern (Athyrium filix-femina) 38
  Western sword fern (Polystichum munitum) 39
Equisetaceae 40
  Common horsetail (Equisetum arvense) 40
Plantaginaceae 41
  Plantain (Plantago) 41
  Veronica (Veronica persica) 42
Ranunculaceae 43
  Creeping buttercup (Ranunculus repens) 43
Onagraceae 44
  Small-flowered willowherb (Epilobium parviflorum) 44
Iridaceae 45
  Douglas iris (Iris douglasiana) 45
Colchicaceae 46
  Fairy bells (Disporum smithii) 46
Asparagaceae 47
  False lily of the valley (Maianthemum dilatatum) 47
  False Solomon's seal (Maianthemum stellatum) 48
Melanthiaceae 49
  Western trillium (Trillium ovatum) 49
Myrsinaceae 50
  Pacific star flower (Trientalis latifolia) 50
Scrophulariaceae 51
  Common figwort (Scrophularia nodosa) 51
Boraginaceae 52
  Forget me not (Myosotis sylvatica) 52
Phrymaceae 53
  Monkey flower (Mimulus) 53
Anacardiaceae 54
  Pacific poison oak (Toxicodendron diversilobum) 54
Sapindaceae 55
  Vine maple (Acer circinatum) 55
Araceae 56
  Western skunk cabbage (Lysichiton americanus) 56
Aristolochiaceae 58
  Western wild ginger (Asarum caudatum) 58
Urticaceae 60
  Wood nettle (Laportea canadensis) 60
Papaveraceae 62
  Pacific Bleeding heart (Dicentra formosa) 62
Violaceae 63
  Evergreen violet (Viola sempervirens) 63

Introduction

Redwood forest is a unique environment and very specific habitat found only in one place on the planet: coast of Northern California and Southern Oregon. Redwood trees are the relic from the old days. They are the tallest trees around. Unfortunately lots of them have been logged and used for industrial purposes so what we have now are designated state parks.

Redwood forest is a rich but very specific habitat that supports a wide variety of life. You will not find a whole lot of different species here but what you find is specific to the redwoods. Dead trees, for example, are home to incredible amount of bacteria that in its turn support abundant plant life.

You will seldom see other species of trees in the redwood forest and the plants that live there are all necessarily will be shade loving or at the very least shade tolerant. One plant that is a signature mark of redwood forest is redwood sorrel: easy to recognize. It grows like a carpet at the feet of the giant trees. Some other plants are seldom seen outside the forest but there are several hardy ones that grow in meadows, roadside, lawns, etc. Redwood forest is a paradise for a botanist and a microbiologist as it offers a wide open field for research and observation.

The plants presented here are mainly from three different location: Arcata community forest: a redwood grove located right in the city of Arcata 7 miles north of Eureka, The Avenue of the Giants—the most unique and interesting redwood grove as it still contains giant trees from the ancient days and a forest near Orick, located approximately in the middle between Eureka and Crescent City in Northern California.

So let's delve in this unique habitat and see what plants we can discover among the silent sentinels.

Cupressaceae
Cupressaceae
Coast redwood (*Sequoia sempervirens*)

These are one of the ancient plants on the planet. The species typically live from 1000 to 2000 years. These are also the tallest trees. Some species reach 380 feet. An estimated 95% of all trees have been cut out. What remained became national and state parks.

Oxalidaceae

Redwood sorrel (*Oxalis oregana*)

Sorrel is the most widespread plant in the redwood forest. It forms a green carpet on the forest floor. Redwood sorrel produces small white to pinkish flowers with purple veins. This plant is related to yellow wood sorrel that decorates countryside and meadows.

Oxalidaceae

Montiaceae

Candy flower (*Claytonia sibirica*)

Candy flower vie with sorrel for the attention. While not as widespread as oxalis, it nevertheless should be noticed. In fact, the flower of claytonia looks very much like that of a redwood sorrel but the leaves are different. It usually hides among sorrel on the redwood floor. Notice cleft petals.

Asteraceae
Asteraceae
Coltsfoot (*Petasites frigidus*)

Coltsfoot is another signature plant of the redwood forest. Its large deeply lobed leaves sure demand attention. The leaves make nice hammocks for banana slugs. When in bloom, it put forward an inflorescence of very small white flowers. The blooming season is short.

Asteraceae

White hawkweed (*Hieracium albiflorum*)

This plant looks a little bit like dandelion and cat's ear though it's usually much taller than them and the flower cup is more pronounced. It is not common. There is also yellow hawksweed.

Asteraceae
Pathfinder (*Adenocaulon bicolor*)

The name refers to a different colour of the leaf if you turn it over. The leaves are green on top and silvery white underneath, so when humans and animals roughly stomp on the plant, the turned leaves mark the trail thus making it easy to find. Pathfinder produces tiny white flowers on a tall stem.

Asteraceae

Nipplewort (*Lapsana communis*)

There are two version why the plant is named nipplewort. The first one refers to small bumps (leaves) on the lower leaves that look like nipples. The other version states that the flower buds look like nipples. The plant is not very common but it's not rare either. In some location it is more abundant than in others. It likes shade.

Asteraceae
Common madia (*Madia elegans*)

Madia is a very rare plant that is easily recognized by the specific shape of the ray florets. The flowers are rather small but not tiny and their colour is yellow.

Brassicaceae

Milkmaids (*Cardamine californica*)

Another uncommon resident of the redwood forest. Milkmaids are usually seen as isolated plants. Look for flowers with four petals that are typical for mustard family though the flowers are bigger than mustard's.

Brassicaceae
Hairy bittercress (*Cardamine hirsuta*)

Hairy bittercress is very easy to miss as it is a very low growing plant and it usually hides among other plants. In addition it is not very common. The leaves are typical for cresses and the white flowers are extremely tiny.

Lamiaceae

Betony (*Stachys*)

Stachys is a common plant not only in meadows and marshes but also in the forest. It has a square fuzzy stem and dark pinkish flowers. Stachys does not have any smell or it can have dusty smell which is uncommon for mints in general. Stachys is noted for its medicinal properties.

Lamiaceae
Common self-heal (*Prunella vulgaris*)

Prunella is a cheerful little flower, very beautiful. It is rather low but when in bloom it puts forward bright purple flowers that are easy to spot and identify. Prunella has a pleasant fragrance when rubbed. As the name implied, the plant is highly regarded for its medicinal properties.

Lamiaceae

Lemon balm (*Melissa officinalis*)

Lemon balm belongs in herb gardens rather than forests and meadows but this plant is very hardy and can handle all kinds of different environments. As a meadow plant it prefers sun but it can tolerate shade as well. The plant is famous for its wonderful fragrance.

Lamiaceae
Field mint (*Mentha arvensis*)

Field mint is the only true mint (*mentha*) from all the plants that we've already discussed. Like other representatives of the mint family, it has a square fuzzy stem, alternating leaves and round flower heads that are pale purple in colour. This mint has a nice aroma and can be used as a spice similar to garden variety mint. The flavour is milder than that of the garden species.

Lamiaceae

Wild mint grows in the forest though only in meadow patches and not in the shade of the trees. Look for square stems and opposite leaves that alternate on the stem. When in bloom the stem of the plant develops pale lavender balls which are actually a cluster of very small flowers. It easily reproduces by vegetative method.

Fabaceae
Fabaceae
Lesser hop trefoil (*Trifolium dubium*)

Another misfit that wandered from the meadows. It is not common but it adds a nice bright colour to the forest meadows. It look similar to more familiar purple and white clovers except this one is smaller and its colour is yellow.

Fabaceae

Sweet pea (*Lathyrus odoratus*)

It was a surprise to see a pea plant in the forest yet it looked wild enough. Peas are actually very hardy plants and they will take over. The bright pink flowers add a cheerful note to the shady habitat.

Rosaceae
Rosaceae
Thimbleberry (*Rubus parviflorus*)

Thimbleberry is related to blackberry, in fact they belong to the same genus. The leaves of thimbleberry are big and deeply lobed and the berries are yellow or orange. They are not as famous for taste as blackberries but ripe thimbleberries are edible and do not taste too bad.

Rosaceae

Rosaceae
Himalayan blackberry (*Rubus armeniacus*)

Himalayan blackberry is an old familiar plant that can be seen everywhere. Some unkind people call it invasive species but we say look who's talking. Himalayan variety and its brother California blackberry (*Rubus ursinus*) can be seen among the shrubs of the forest.

Rosaceae

Rosaceae
Salmonberry (*Rubus spectabilis*)

Another rubus genus. Salmonberry is not as common like its famous brothers blackberry and thimbleberry but redwood and other kind of forest is a good place to look for it.

Berberidaceae

Deerfoot (*Achlys triphylla*)

Deerfoot or vanilla leaf is a rare plant in the redwood forest though it seems it only grows in the forest. It has big leaves divided into three part that look almost like separate leaves but they are connected. Contrary to what the common name implies the plant does not smell like vanilla when dry.

Berberidaceae
Inside-out flower (*Vancouveria hexandra*)

Vancouverua is a very unusual plant. As the name implies, it does look like the flower is attached to the wrong end of the stem. The leaves look a bit similar to that of a meadow rue. The blooming period is short but the leaves can be seen even when the plant is not blooming.

Berberidaceae

Oregon grape (*Mahonia aquifolium*)

Oregon grape is typically a bush though quite often the new growth is seen as a low plant. It is not related to true grapes (*Vitis*) from which wine is produced but rather because it produces fruits or berries that look a little bit like blue grapes. The berries are rather small. The leaves have a distinct shape that allows for easy plant identification.

Apiaceae
Apiaceae
Pacific sanicle (*Sanicula crassicaulis*)

Sanicle is rather tall plant with flower heads (umbrellas) consisting of tiny yellow flowers. Unlike its more widespread cousins wild carrot or water parsley, sanicle's umbrella is rather small and the plant itself is quite inconspicuous. The leaves are dark green, glossy and deeply lobed.

Apiaceae

Water parsley (*Oenanthe sarmentosa*)

Water parsley is one of the most ubiquitous plants around. It is usually found in meadows but it feels quite comfortable in redwood forest as well. When in bloom water parsley produces an umbrella of tiny white flowers. The leaves are edible.

Grossulariaceae
Grossulariaceae
Currant (*Ribes*)

There are two currants that grow in our forest: red currant (*Ribes rubrum*) and black currant (*Ribes nigrum*). Besides currant the *Ribes* genus includes gooseberries (*Ribes uva-crispa*). The plant is not encountered very often.

Ericaceae

Huckleberry (*Vaccinium*)

Like currants, there are two varieties of huckleberries: red (*Vaccinium parvifolium*) and black (*Vaccinium resinosum*). Black huckleberry is widespread while the red variety is rare. They both produce berries of the aforementioned colours. The leaves of red variety are smooth while the ones of the black vaccinium are serrated.

Saxifragaceae
Saxifragaceae
Alumroot (*Heuchera*)

Alumroot grows in abundance in some places while being rare in others. Heuchera is rare in redwood forest and it can esily be confused with piggyback plant. The leaves look like that of cultivated geranium. The tiny white flowers grow on tall stem.

Saxifragaceae

Piggyback plant (*Tolmiea menziesii*)

Tolmeiea looks very close to heuchera, in fact, they are very hard to tell apart as the shape of the leaves is similar. Piggyback plant has tiny leaves growing at the root of the petioles, hence the common name and the identifying feature. Its blooming season is very short.

Dryopteridaceae
Dryopteridaceae
Lady fern (*Athyrium filix-femina*)

Ferns. The signature plants of the forest. Ferns are associatd with real jungles and pristine forests. Redwood forest has them too. The most abundant species are lady fern and sword fern, both are easy to recognize by the shape of their leaves. Unlike flowering plants, ferns propagate by spores. The spore pockets are situated on the underneath of the leaves.

Dryopteridaceae
Western sword fern (*Polystichum munitum*)

Sword fern is extremely common and grows especially well in shady areas. The leaves look like swords and the lower area is a little bit narrower than the middle one. The top area is narrower than the middle and the bottom ones. Sword fern looks very similar to bracken fern but unlike the latter is not edible.

Equisetaceae
Equisetaceae
Common horsetail (*Equisetum arvense*)

Common horsetail is another plant that propagates through spores. It is highly regarded for its medicinal properties. Horsetail is supposed to be edible but special preparation is required.

Plantaginaceae

Plantain (*Plantago*)

Plantain is a very common plant that grows well in the sun or in the shade. The young leaves are rumoured to be edible. The leaves of mature plant are too fibrous to be any good. The two most common species are broadleaf plantain (*Plantago major*) and English or narrowleaf plantain (*Plantago lanceolata*). Plantain flowers are pollinated by wind.

Plantaginaceae
Veronica (*Veronica persica*)

Veronica is a low growing herb with small attractive blue flowers that have a peculiar shape allowing for easy identification. Common names include Persian speedwell, winter speedwell or bird's eye. There are lots of species of veronica.

Ranunculaceae

Creeping buttercup (*Ranunculus repens*)

Creeping buttercup is the most common wild plant in Ranunculus family. This buttercup can be found in all kinds of habitat like meadows, forests and even roadside. The plant has bright yellow attractive flowers and dark green deeply lobed leaves. The leaves can still be seen after the bloom is gone.

Onagraceae
Onagraceae
Small-flowered willowherb (*Epilobium parviflorum*)

Willowherb is another common plant that can be seen everywhere including backyards and curbs. It has small (*parviflorum* means small-flowered) pink or purple flowers. The peculiar feature of the plant is that each of its four petals are cleft to the middle of the length of the petal.

Iridaceae

Douglas iris (*Iris douglasiana*)

Douglas iris is a rare plant in the redwood forest. Irises have beautiful flowers that add blue splash of color to the dark greenery of the grass. Irises have peculiar flower structure but it varies in some genera.

Colchicaceae
Colchicaceae
Fairy bells (*Disporum smithii*)

Disporum look like little lilies and in fact it used to be part of the lily family but recent phylogenetic studies based on molecular research indicated that lots of genera previously considered part of the Liliaceae should be classified as separate families. Disporum is only one such case.

Asparagaceae

False lily of the valley (*Maianthemum dilatatum*)

Like the name and appearance indicates, this plant is not a true lily and now is not even considered part of the lily family. When the plant matures, it produces nice bright smooth blotched berries.

Asparagaceae
False Solomon's seal (*Maianthemum stellatum*)

False Solomon's seal (or star-flowered Solomon's seal) is a very attractive plant. The leaves are smooth and shiny with parallel veins. The flowers are white, very small that eventually turn into berries. Sometimes you can see meadows of this plant among the redwoods. The young shoots (not the leaves) are edible. The plant also possesses medicinal properties.

Melanthiaceae

Western trillium (*Trillium ovatum*)

Western trillium is a beautiful plant that is specific to the forest habitat. It grows well in the shade but it requires partial sun to bloom. The mature plant produces seeds that are spread around by ants.

Myrsinaceae
Myrsinaceae
Pacific star flower (*Trientalis latifolia*)

This is a low growing herb with beautiful flowers that indeed look like little stars. The leaves grow in rosette on the upper part of the stem. There is usually one flower per stem. The plant is small and is easy to miss.
According to some botanical nomenclatures, the plant belongs to the family *Primulaceae*.

Scrophulariaceae

Common figwort (*Scrophularia nodosa*)

Figwort is a worthy representative of the figwort family. It is an attractive plant with numerous tiny red flowers when in bloom. The plant distribution is wide and it can be found in all kinds of habitat and climate ranges. When not in bloom, it can be confused with stachys and nettle.

Boraginaceae
Boraginaceae
Forget me not (*Myosotis sylvatica*)

Myosotis is a low growing plant that produces small flowers, usually blue. The interesting thing about it is that in other European languages the folk name is still translated like forget me not. The leaves and stems are fuzzy, which is typical for the members of borage family.

Phrymaceae

Monkey flower (*Mimulus*)

Monkey flower can be confused with penstemon but it is actually a totally different plant, they don't even belong to the same family. Though they both used to be placed in the family *Scrophulariaceae* as morphologically these plants look close to each other. The new placement is based on modern genetic research. There are numerous species of monkey flower genus.

Anacardiaceae

Pacific poison oak (*Toxicodendron diversilobum*)

There are two varieties of poison oak: Pacific (*Toxicodendron diversilobum* or *Rhus diversiloba*) and Atlantic variety (*Toxicodendron or Rhus pubescens*). Both belong to the same genus which is part of sumac or cashew family. This plant is famous for releasing substance that on contact with the skin can cause rash and painful feeling. The plant varies in appearance and the leaves look similar to oaks, both European and California varieties.

Sapindaceae

Vine maple (*Acer circinatum*)

This plant belongs to the same genus as regular sugar maple (*Acer saccharum*). Unlike sugar maple, the vine maple grows in a shrub, seldom into a small tree. The fruits are two seeded samaras.

Araceae

Western skunk cabbage (*Lysichiton americanus*)

Skunk cabbage or swamp lantern is easily recognized by its big leaves and specific habitat: it only grows close to the streams. The leaves are very big and sturdy. The flowers form a spike that is covered by yellow hood called spathe. When the plant matures, it sheds the hood. It also has characteristic unpleasant smell of a skunk. This odour attracts pollinators like flies and beetles. The plant is eaten by bears when they come out of hibernation in the early spring. This plant was also used for medicinal purposed by American Indians.

Araceae

Aristolochiaceae

Western wild ginger (*Asarum caudatum*)

One of the typical plants of the forest that prefers shade. The leaves are round, non glossy and heart-shaped with spread out veins. The flowers are usually hidden under the leaves but the blooming season is very short. The plant is supposed to smell like ginger when rubbed but it does not.

Aristolochiaceae

Urticaceae

Urticaceae

Wood nettle (*Laportea canadensis*)

Wood nettle can be confused with other plants, namely stachys and figwort when not in bloom as the leaves of all three species look very similar and all three have square stem. But the blooms of these plants are very different. Nettle have tiny flowers forming tassels. This mechanism developed so they could be pollinated by wind.

Laporea is not a true nettle in a botanical sense, those are part of the genus Urtica. However stinging hairs are still present, which gives the latter genus and the family its name. The genus name Urtica come from the Latin word *urere*, which means "to burn".

Dock leaves (*Rumex*) are supposed to relieve the painful feeling of sting. I was stung by a nettle once and grabbed a leaf of star-flowered Solomon's seal, rubbed it and applied the bruised part to my finger. The pain was gone almost immediately.

Urticaceae

Papaveraceae
Papaveraceae
Pacific Bleeding heart (*Dicentra formosa*)

Bleeding heart is a beautiful plant (formosa means beautiful) with pink flowers that are shaped like a heart. It is usually found in flower beds but can be seen in a redwood forest.

Violaceae

Evergreen violet (*Viola sempervirens*)

In spite of the name this flower is yellow with dark stripes on the lower petal. It is one of the earlier blooming plants in the redwood forest though not quite as common as redwood sorrel. Violets have round heart-shaped leaves.

Another violet that can be seen in the wild is western dog violet (Viola adunca). Now that violet is violet that is to say the color of the flower is violet.

Made in the USA
Las Vegas, NV
29 March 2025